Norman
…A Most Unforgettable Character

by

Capt. Jim Blackburn

PublishAmerica
Baltimore

First printing

ISBN: 1-4241-7058-3
PUBLISHED BY PUBLISHAMERICA, LLLP
www.publishamerica.com
Baltimore

Printed in the United States of America

To Norman's wife, Edna

Acknowledgments

Thanks to the following for their help:

Edna Riddell

Helen Diaz (Norman's daughter)

Dennis M. Giangreco, author, *Airbridge to Britain*

Contents

Introduction

It was in early winter of 1983 when I first met Norman Riddell. Eastern Air Lines had taken over the South American routes of bankrupt Braniff Airways in 1982. As a captain with Eastern, I was flying to our newly acquired destinations. This was my first flight to Santiago, Chile. Norman's daughter, Helen, was now one of our flight attendants, having previously flown for Braniff. As a courtesy to her crewmates, she told us about the British pub, Cross Keys, that her parents owned and operated in Santiago. She assured us that we would find a warm welcome there, and she was right about that. Her parents would give us advice on where to find the best restaurants, shops and even give us the best exchange rate for our dollars. It was on this visit to Santiago that I first met Norman and Edna Riddell. Over the next five years, I bid flights to Santiago as much as my seniority allowed, as it was my favorite layover spot in South America. A great deal of my attraction to this large metropolitan city was the wonderful friendship that developed between Norman and me. In our earlier flying days, we had both been military bomber and transport pilots. The great fraternity of aviation, with its many shared experiences, provided us with topics of discussion to last a lifetime. Many people had urged Norman to write down details of his many adventures, but he had always declined. He said that he didn't want to appear to be bragging. On several occasions, while visiting with Norman, I copied down some data on his exciting life and will attempt to give you an overview of it here, in the following narrative.

Chapter 1: The Early Days

Norman Charles Stewart Riddell was born on October 16, 1916, in Thames Ditton, Surry, England. This was a turbulent time, as World War I was raging. Norman's father (Norman, Sr.) was a bomber pilot in the Royal Flying Corps at that time. He had been a boarding student from Chile at the University of Edinburgh, in Scotland, when the war broke out in 1914. He enlisted in a Scottish cavalry regiment, but later transferred to the Royal Flying Corps and became a pilot. It was in Scotland where he met his future wife, Elinore Stewart Neugent. They married in 1915 and a year later, Norman, Jr., was born. After the war, in 1919, Norman's parents moved to the South American nation of Chile, where the Riddell family had lived for generations.

NOTE: Norman's great-grandparents had first moved to the country of Chile in about 1855. They were in the ships chandler business in Valparaiso. Later, Norman's grandfather opened a big department store there and was very successful. The Riddells, and their generations to follow, would go back to Britain for the birth of their children, but always eventually returned to their adopted Chilean home.

Norman was given a rather lavish education. He attended boarding schools in both Scotland and England. It was about 1923 when Norman was sent from Chile to Scotland to attend the Dollar Academy (established in 1818). This is the oldest co-educational boarding school in the world. A few years later he became a boarding student at the famous Blundell's School (Established in 1604) in Devonshire, England.

During Norman's early days at the boarding schools, he found out that boxing lessons were available to any young lad who was interested in learning to, as Norman would say, "put up your dukes." He became quite proficient in this sport and it served him very well in years to come. A huge problem arose when the Great Depression came about in 1929. Norman's family could no longer get funds to England for his tuition and he, and many other students, had to leave school and return home. After his return to Chile, he was one of the first students to attend the very exclusive Grange School. This school was to become one of the best private schools in the country. Norman had a strong body, fine mind and quick wit, which lasted him his whole adventure-filled life.

Chapter 2: A Love of Flying

In 1930, during the days of the Great Depression, Norman was a young lad of 14. He got a part-time job with the Cable & Wireless Company, acting as a messenger and translator for the president of Chile. It was about this time that Norman developed a great yearning to learn to fly an airplane. Norman's dad had turned down his request to teach him how to fly, so Norman would visit the El Bosque Airport, near Santiago, and do odd jobs for the pilots there. In appreciation, they would give him a little pay, but, more importantly, *flying lessons*. Norman was a quick learner and very big for his age. After learning to fly, he got a job with a company that carried the mail over the Andes Mountains to Argentina. He told one story about having a tire that burst upon landing in Mendoza, Argentina. There was no equipment to repair the damage, so Norman stuffed the tire with cow dung he obtained from a nearby pasture and continued on his journey. Ever resourceful, Norman loved flying and became quite good at it. He was a natural.

In 1932, Norman secured a post as a copilot with Panagra Airways. Panagra was owned jointly by Pan American Airways and the W. R. Grace Steamship Company. Grace operated a fleet of freighters and ocean liners, sailing to South America. Norman was a big strapping young man with a record of good experience flying in the Andes, and Panagra considered themselves fortunate to be able to hire him. He flew up and down the West Coast of South America to the United States, carrying passengers and international mail. Flying through the Andes was very dangerous because of the bad weather and high terrain, but Norman thrived on it and gained much valuable experience.

In 1933, Norman was sent to San Antonio, Texas, to get checked out on a new type airplane. It was there that they discovered, after closely checking his file, that he was only 17 years old. Juan Trippe, the president of Pan American and Panagra Airways, told him that he was still a year too young to be employed as a pilot with Panagra. The minimum age was 18. Because of his tall, muscular build, the Panagra office personnel in Santiago had failed to notice his date of birth, and consequently his young age. Norman said that, as a consolation, they graciously sent him back home to Chile on a Grace passenger liner, with all expenses paid. Norman may hold the record, at age 16, for being the youngest airline pilot ever hired. Back in Santiago, Norman served as a voluntary fireman in Chile's fire services, but still yearned to fly more than anything else. Biding his time, he went back to work for the Cable & Wireless Company for a few years.

Chapter 3: RAF—Adventure Calls

In 1936, at the age of 19, Norman returned to England to apply for training as a junior flight officer in the RAF. During the time he waited for acceptance to RAF flight training, Norman looked around for something interesting to do. His love of flying and adventure took him to Barcelona, Spain. There he joined the Spanish Republican Army, to participate in the Spanish Civil War. He really wanted to fly, but when the army recruiters asked him where he was from, he said, "Chile," and they said, "Cavalry…go steal a horse." He was involved in some heavy fighting in the Madrid area, and said it was sheer hell. After a few months he left the bloody battles, made his way to France and then on back to England. There he found that he was to receive the best Christmas present he could possibly hope for, that of being accepted into the RAF flight-training program as a cadet.

It was not easy for an applicant to be accepted into the RAF for pilot training in early 1937. The RAF, at that time, was looked upon more as a gentlemen's flying club than a branch of the military establishment. Norman said that he had a leg up in getting accepted, primarily because of his father's service as a pilot in the Royal Flying Corps in World War I.

Norman went through pilot school at RAF Sywell, Northants, and at No. 5 Flight Training Squadron, Sealand, Cheshire. As a very junior officer in flight training, Norman found that the pay was a pittance. After paying mess bills, there was hardly any money left for beer and fun. Word soon got around though, that Norman was quite good at boxing. He and his pals decided on a scheme to make a little pocket money. This led them to the local area county fairgrounds, where one of the attractions was a boxing ring. There was a big gorilla of a man there, waiting for someone to challenge him in the ring. A cash prize was given for anyone who could last a few rounds with this hulk of a man. Needless to say, Norman's pals quickly had him in the ring, taking up the challenge. To their relief, Norman lasted the required number of rounds and came away none the worse for wear…not even a bloody nose. Norman was able to dance circles around the big brute of an opponent and his boxing expertise and husky, 20-year-old body served him well. Norman's chums had cheered him on while he was in the ring and the boxing prize money was now theirs. They hoisted him on their shoulders and headed for a local pub to celebrate Norman's victory, singing, "For He's a Jolly Good Fellow." This little escapade continued about once a month, each time with fruitful results and no shortage of beer, but, alas, all good things come to an end. One night when Norman and his pals descended upon the fairgrounds boxing ring, the owner spotted their arrival. He begged them not to do him in again, and said that they were ruining his business. Then he just gave them the ring prize money and pleaded with them to never come back again. Sad to say, but that was the end of their little scheme to make a few pounds on the side and keep their beer and spirits flowing.

In 1937, Norman was initially assigned to long-range bombers in Winchester, Hampshire, and completed his flight training there.

In 1938, Norman was assigned as a flying officer to No. 207 Bomber Squadron in Cottesmore, Rutland, flying the Fairey Battle light bombers. As war clouds gathered over Europe, the RAF was preparing for the inevitable conflict to come.

On September 5, 1939, Britain and France declared war on Germany in response to the invasion of Poland. There had been no invasion of France by the Germans early on, but it was to come later, with a vengeance, in a spring offensive. Norman joined No. 88 Squadron as a pilot officer on Fairey Battle bombers at Boscombe Down, and moved to France, based at an airfield near Reims. Bomber Command had sent ten squadrons of Fairey Battle bombers into France to serve in an Advanced Air Striking Force, to be the first line of defense against any Nazi offensive. From his base in France, Norman wrote to his father to say that he was with 88 Squadron. To his surprise, his father replied saying that he, too, served as a pilot with No. 88 in 1917, at an airfield near Arras, in northern France. This was during his WWI service in the Royal Flying Corps.

Norman said that the winter of 1939-40 was extremely severe and it was difficult to keep their bombers operational in the heavy ice and snow. They played a "waiting game" for the expected German offensive to begin. This period of time became known as the "Phony War." War had been declared, but neither side had yet to start an offensive.

Flight of Norman's 207 Squadron, Fairey Battle Bombers in 1938

Chapter 4: World War II Combat Begins

In early May, 1940, the German land offensive finally began. The Nazi "Blitzkrieg" rolled rapidly into Belgium and Norman flew many sorties in the Fairey Battle light bombers against the German onslaught. Over half of his squadron mates were shot down as they valiantly tried to hold back the German advance towards Brussels, Belgium. The British Expeditionary Force ground troop survivors had to be evacuated from Dunkirk. With what was left of Norman's bomber squadron, they were given the task of flying night attacks on German lines of communication. Because of the ferocity of the German onslaught, Norman flew a lot more than the required number of missions and became totally exhausted. Just before the fall of France, he was finally ordered back to England. On June 17, 1940, he found himself regrouped at RAF Sydenham, near Belfast, Ireland. There, it was determined that Norman was suffering from battle fatigue and he was put in a military hospital. Norman called this hospital a real "loonie bin" and, after a few frustrating weeks of idiotic tests, convinced them that he was not crazy. He was then returned to RAF active duty.

In July 1940, a month after returning from France, Norman was given a challenge. He was to be the first RAF pilot to fly the Douglas A-20A light attack bomber, but first it had to be assembled. Just prior to the outbreak of WW II, the French government had placed an order with Douglas Aircraft (USA) for a consignment of planes. These were shipped in crates as deck cargo on board merchant vessels. France had received a few, but had fallen to the Nazis, and the later plane deliveries were diverted and offloaded in Ireland. These planes were transferred to Norman's new station at RAF Sydenham. Also based there was an aircraft engineer by the name of "Tex" Nixon. He was an employee of Lockheed, but was on loan to Douglas Aircraft at that time.

Tex approached Norman and informed him that when he had assembled one of those A-20A aircraft, he would like Norman to fly it, and would go along with him. Norman asked him if he realized what he was letting himself in for, because he had never had any previous experience flying as a lone pilot on a twin-engine aircraft, furthermore one with a tricycle undercarriage and reversed throttles.

NOTE: The French contract had specified that their A-20A aircraft have throttles that increased power as you pulled them back, not as you pushed them forward. This was just the opposite rigging of most other nation's planes, but standard, at the time, in those of the French Air Force.

Since this was a single-pilot plane, Tex had to lie down over the bomb bay, just behind Norman's pilot seat during flight, so he could answer any questions Norman may have. The flight test went well, but it was determined that the A-20A was grossly under-powered. They recommended equipping the plane with the more powerful Pratt & Whitney Cyclone engines. This was accomplished later, along with adding armor plating, self-sealing fuel tanks, and re-rigging the throttles to normal British specifications. It became known in the RAF as the Douglas "Boston" bomber. (The upgraded A-20 was supplied later to RAF No. 88 Squadron, in November of 1941).

Chapter 5: The Battle of Britain

In August 1940, while Norman was completing the successful test flying of the A-20 bomber, Britain had come under the massive assault of fighters and bombers of the German Luftwaffe. This was the beginning of the great air war known as the "Battle of Britain." The RAF Fighter Command was pressed to their limit to supply planes and pilots for the battle. Due to the early losses, Norman was "loaned" to Fighter Command and posted to No. 242 Squadron, flying Hawker Hurricane fighter planes. His squadron operated from a remote auxiliary airfield, located away from most of the main targeted military and industrial areas of England. Many huge air battles took place over Britain in the next few months. During his quick training to fly the Hawker Hurricane Fighter in combat, Norman said that he was told by his instructor to remember a couple of cardinal rules. *"When in combat, don't fly in any one direction for very long, and always turn and face the attack."*

In one of his many air battles, Norman said that he noticed a German ME-110 in the rearview mirror of his Hurricane. Just as he was deciding on which direction to break, he was hit with a burst of machine-gun fire, and wounded in both legs. He was also wounded in the face when part of his canopy shattered. Norman did manage to evade the enemy fighter and was able to make a successful emergency landing on the Isle of Wight. RAF fighter pilots had been encouraged to bring their damaged aircraft back and land, if at all possible. The ground crews were experts at repairing and rebuilding damaged planes, even if it was only to use them for the vital spare parts.

Norman was hospitalized for a number of weeks at East Grinstead Hospital, where he received good care and excellent plastic surgery.

Later, after discharge from the hospital, he was posted to Keflavik, Iceland, with RAF No. 98 Bomber Squadron. While he recuperated there, he participated in a number of experimental RAF programs. One involved a smoke screen test, and another had to do with balloon cable cutters.

NOTE: The RAF lost 832 fighter aircraft in the Battle of Britain, but many pilots survived, to fly more missions. They shot down 668 German fighters and over 600 bombers, which caused the German Nazi leader, Adolph Hitler, to put his September 1940 invasion plans on hold.

Chapter 6: Battle for Singapore

In early 1941, Norman was assigned to No. 21, South African Squadron, flying Beauforts and Marylands, and participated in bombing the Nazi-controlled Vichy French fleet in northern Madagascar. They flew support missions in the bombing of the island to help secure the vast supply of graphite, a strategic war materiel. From there he was assigned to No. 84 Bomber Squadron at RAF Dum-Dum Airdrome in Calcutta, India, where he flew Bristol Blenheim Mark IV bombers.

In late 1941, Norman's 84 Squadron was sent to Palembang Airfield on the island of Sumatra. Because of the rapid advance of Japanese invasion forces, Norman received orders to be "loaned" to Fighter Command once again. This time he was to be sent on temporary duty (TDY) to "Fortress Singapore" to bolster the island's fighter defense. He was to instruct the novice Hawker Hurricane and Brewster Buffalo fighter pilots in aerial combat tactics, using the experience he had gained in the Battle of Britain. Most of the RAF pilots in Singapore had never flown in combat, but their Japanese enemy was very well experienced.

NOTE: The British had built a big naval base on Singapore that was finished in 1938. Its large naval guns were meant to hold off an invasion from the sea. The guns (the Jahore Battery) could be swiveled around 360 degrees, but only had armor-piercing shells made for sinking warships. They were unsuitable for use against land combat troops. The British commander, General Sir Arthur Percival, failed to provide defenses to repel an invasion from the Malay Peninsula and northwest channel. The general had been ordered to build defenses in these areas several times, but refused because he said, "it would be bad for morale." This phrase would come back to haunt him later.

By a strange quirk of fate, the German surface raider Atlantis *had shelled and stopped a British cargo ship near Sumatra in November of 1940, nearly a year earlier. The German boarding party had broken open a strong-room on the ship and discovered a treasure trove of British documents. This included the Top Secret mail for the British High Command; new code tables, and a War Cabinet report on British forces,* Defenses of Singapore. *After its crew and passengers had been taken prisoner and this vital intelligence data had been removed, the ship was sunk. It was not known until years after the war that the Germans, who had shared the data with the Japanese, had captured these vital documents in November of 1940. This, coupled with having spies on Singapore Island, played a great part in the Japanese decision to launch their East Asia War.*

This was the situation into which Norman was pressed. The Japanese had experienced combat pilots, air superiority and more modern planes, but the RAF pilots on Singapore did their best with what little they had. When the Japanese invasion came over land, they made rapid progress into Singapore. Just a few days before the surrender, Norman got some other pilots together and developed a plan to evacuate, rather than be taken prisoner. Norman led a flight of the last serviceable, U.S.-built, Brewster Buffalo fighter planes out of Singapore. They made a series of daring flights in one- to two-hour hops to RAF auxiliary airfields in Sumatra, Thailand, Burma and, finally, to RAF Dum-Dum Airfield in Calcutta, India. During this perilous journey, Norman would land first at each of the auxiliary fields while his flight mates circled above. This was a precaution in case Japanese troops had captured the airstrip. (The Japanese had coerced the Thai government into an alliance and had stationed troops in some parts of Thailand.) On the ground, Norman's pilots had to refuel their planes by hand, pumping gas from steel drums. Meanwhile, back in Singapore, General Percival raised the white flag and surrendered over 130,000 troops on February 15, 1942, little knowing that the Japanese were nearly out of ammunition.

Brewster Buffalo Fighter—
photo courtesy of British Aircraft of World War II.
(Similar to ones flown by Norman and his flight mates,
escaping from Singapore)

Back at RAF Dum-Dum in India, Norman and his fellow pilots were threatened with being brought up on charges of dereliction of duty, by a superior officer. The idea of charges were dropped after Norman explained that they had done all they could to defend Singapore, and could see no good purpose in becoming prisoners of war. They were actually, as Norman put it, "conserving assets of the Crown," and living to fight again another day. Score another one for the quick-thinking Norman.

Chapter 7: Norman's Iraq Adventure

In 1942, as the war progressed, Norman was first sent to Iran, and then on to Iraq, on the Persian Gulf. Iraq had once again, as now, become a source of trouble. There was the Anglo-Iraqi Treaty of 1930 that gave the British permanent rights to two airfields in Iraq. One in the south, near Basra, and the other, a desert flight-training school, near Fallujah at Habbaniyah. Free rights of access and transit were guaranteed to the British by this treaty.

NOTE: In 1941, Rashid Ali, who was working with the Germans, had become prime minister. He staged a revolt and attempted, along with three prominent Iraqi officers, to militarily force the British out of Iraq. Rashid Ali called this the "Arab Freedom Movement," but it was just a ploy to deliver Iraq into German control. With great daring and a minimal force, the Brits had subdued these rebel troops and kept the Germans from taking over the Iraqi oil fields for their Nazi war effort.

Norman was once again "loaned out." This time it was to the "liberated" Iraqi Air Force, and he was based in Basra. While there, one of the Arab leaders asked Norman to teach his son to fly. This was a real challenge, because the leader's son had never even driven a car, and was severely lacking of any talent in the field of aviation. Norman did his best, but said that the "Prince" was not destined to become a pilot of any great distinction, and "bloody lucky if he didn't kill himself in the learning process."

Norman noticed a strange thing about his paychecks while being based in Iraq. He said his checks were almost double the RAF pay he had been getting previously. He was delighted with the increased pay, but thought that it may have been a mistake, and if it was, didn't want to have to repay a lot of money at a later date. He made inquiries and found out that, as RAF liaison officer, he was being paid by the Iraqi Air Force, and at the officer grade *they* determined his position called for. Not to argue with a good thing, Norman enjoyed the extra money and praised the Iraqi Air Force leaders for their insight into his true worth.

Chapter 8: Croix de Guerre

In 1943, Norman left the Iraqis and joined No. 52 Squadron in the 1st Tactical Air Force (Desert Air Force), where he remained all through the North Africa, Sicily and Italy campaigns.

In May 1944, Norman returned to Britain and immediately rejoined No. 98 Squadron at RAF Dunsfold, on B-25 "Mitchell" medium bombers. While there, he met and became friends with U.S. Air Force General Jimmy Doolittle. Norman said that they had met on four occasions during the war and become close friends. He flew many bombing sorties over France, Belgium and into the border area of Germany. In September his squadron moved to a captured airfield near Brussels, Belgium. Then it was on to Holland at Melsbroek with the 98th Squadron, still flying the B-25s.

The liberation of France from German occupation, in 1944, gave Norman the opportunity of personally piloting French General Charles de Gaulle back and forth to France. This established a strong bond of trust between the two men. So much so that on de Gaulle's triumphant return to Paris on August 26, 1944, Norman was requested to personally accompany General de Gaulle, to be by his side, as he marched through the Arch de Triumph, down the Champs Elysees, and into the heart of the city. Norman was not permitted to wear his RAF uniform in this parade, however. He had to quickly dress in a borrowed, ill-fitting civilian suit. The arms and legs of the suit were a bit too short, but Norman obliged the general with honor. Norman was pictured, marching with General de Gaulle, on the cover of *Time* magazine. Norman was awarded the French Croix de Guerre medal for his services to France. He also was awarded the British DSO and DFC medals, among others, for his gallant flying exploits in the RAF.

In January 1945, Norman left 98 Squadron, and on return to Britain, was posted as an inspector at the Examining Squadron of the elite Empire Central Flying School. This was a top assignment for an RAF pilot. While there, he was able to fly many different types of current and experimental aircraft. This included the early jets, such as the Gloster Meteor fighters.

Chapter 9: BSAA Airline Days

In June 1946, Norman was talked into retiring from the RAF after ten years of faithful service and join the state-run British South American Airways. BSAA was founded by RAF Air Vice Marshal Don Bennett, a close friend of Norman's in the RAF. AVM Bennett knew that Norman had lived in Chile and was familiar with flying in the South American Andes. He thought, correctly, that Norman would be a great asset to BSAA. They began transatlantic services in March 1946, with Avro York passenger aircraft. They flew the first international flight out of the new Heathrow Airport. Norman flew as captain on Avro Yorks and Lancastrians (Lancaster bombers, converted for carrying passengers), but spent most of his time flying the new Avro Tudor aircraft.

NOTE: BSAA's North Atlantic Route was flown via Lisbon, Portugal, on to Santa Maria Island, in the Azores, across to Bermuda, down to Kingston, Jamaica, West Indies, and on to the west coast of South America, down to Santiago, Chile, and then east, across the Andes, to Buenos Aires, Argentina. Norman was finally able to get back again to the familiar territory of his early flying days and childhood adventures.

In 1947, Norman was enjoying flying as an airline captain with BSAA and later was made station commander at Santa Maria, in the Azores.

It was in August of 1947 that BSAA lost a Lancastrian aircraft. This plane just seemed to disappear into thin air. The plane, named *Stardust*, was flying from Buenos Aires, Argentina, to Santiago, Chile. This was, at the time, a flight of about four hours. Just before Stardust was due to arrive in Santiago, its crew sent out a Morse code message confirming their arrival time, but ending with a strange series of letters. This really baffled the radio operator in Santiago, and, after that, nothing more was ever heard from *Stardust*. It was the start of one of the greatest lingering aviation mysteries.

It wasn't until 53 years later that parts of the plane began to appear from the lower portion of the Tupangato Glacier. This was *not* the crash site. *Stardust* had hit the mountain near its peak and the glacier ice and snow had covered it. It became part of the glacier and gradually, over the years, slid down the mountain, until the temperatures at the lower altitudes caused the ice to melt and expel parts of the crashed plane.

Norman knew most of the crew, because BSAA was a relatively small airline. He remembered flying some of the search flights over the Andes, with no trace seen of the missing plane. There was a lack of radio beacons and other navigational aides in 1947 and a probable cause of the accident was reached years later. Analysis of the old weather charts showed that, on the day of the crash, *Stardust* was flying directly into a jet stream, which was blowing about 100 miles per hour at that time. This was a weather phenomenon that was unknown at the time. A solid cloud deck obscured the tops of the highest of the Andes Mountains, thus there was no visible ground reference. The crew had no way of knowing that a high-altitude jet stream was slowing them down, destroying all their navigational calculations. Nearing Mendoza, Argentina, just prior to crossing the Andes, *Stardust* had climbed to 24,000 feet. This was about 1,500 feet higher than the Andes' highest peak, Aconcagua. With no ground reference, the crew calculated that they had passed over the Andes and had started their descent through the clouds for landing at Santiago. Unfortunately, due to the strong headwinds, they were still some 50 miles east of their assumed position, and right over the Mount Tupangato glacier. The reason that Norman and the other search pilots never saw any wreckage was probably due to the plane being covered with ice and snow just after the crash.

NOTE: Two more BSAA aircraft were lost, over the Atlantic Ocean, and no trace of them was ever found. A Tudor aircraft disappeared while flying between Santa Maria, Azores, and Bermuda in January of 1948. Another Tudor vanished between Bermuda and Kingston, Jamaica, in January of 1949. At the time, many speculated that the curse of the Bermuda Triangle had struck again. Norman didn't think so. He suspected structural failures in the tail of the Tudor planes.

Two years later, in 1951, a movie, *No Highway in the Sky*, was made about a fictional British transport plane that developed structural tail problems. Jimmy Stewart had the male lead in the movie. The fictional airplane depicted in the movie had a large tail, similar to the United States-made Lockheed Constellation. The Avro Tudor plane had a very large vertical fin, which was later modified after the plane developed vibration and performance problems. The Tudor also had problems with cross-wind landings. In those days, wind-tunnel testing had not yet been perfected and computer-generated models didn't exist. Following the mysterious loss of the second Tudor over the Atlantic Ocean, the Tudor was grounded. On July 30, 1949, BSAA was officially merged with BOAC (British Overseas Airline Corporation).

Chapter 10: Berlin Air Lift Operation

In 1948, Norman met the love of his life, Edna, through his younger brother in London. Edna says that she was not impressed with Norman on those first few dates, but later began to really appreciate his personality and wit. Their relationship progressed so well that they were later married and started to raise a family.

Shortly thereafter, Norman was called upon to fly into Germany once more. In July 1948, the Russians had shut down all ground travel into and out of the allied zones of Berlin. The city of Berlin was located deep in the center of the Russian Zone of Occupied Germany, and divided into four zones....American, British, French and Russian. The Americans and British had set up a joint zone in West Germany called "Bizonia," and made plans to issue new currency in both West Germany and West Berlin. This would effectively shut down the black markets and stabilize the West German government. The Russians didn't want an independent West Germany. They wanted it *all* under their Communist control. In retaliation, they were trying to drive the allied forces from Berlin by a blockade. They thought that, without supplies, all of West Berlin and its population would fall under their control. The only way in or out of West Berlin then, was by *air*.

The U.S. Air Force came up with a plan to try to supply the survival needs of the West Berlin by the greatest feat of aviation ever attempted. The solution was the *Berlin Airlift*. Thousands of aircraft were assembled from all over the allied world to participate. The USAF primarily provided Douglas C-54 transports. (I flew several of these planes a few years later, and they still had coal dust in the bottom of the fuselages.)

The British provided about 40 Avro York transports from the RAF, plus some contracted civilian airline planes to haul cargo. Norman flew the British South American Airways (BSAA) Avro Tudor aircraft in the airlift. At the peak, during the later part of 1948 and early 1949, there was one landing and takeoff at West Berlin airports every three minutes. The impossible was being done, but at the tragic cost of many lives. Forty-one British pilots alone were lost in the Berlin Airlift. Norman survived the heavy and dangerous flight schedule, but began having trouble with his legs. Norman continued to fly the Berlin Airlift, but the wounds he had suffered in the Battle of Britain in 1940 began to flare up again. (He had been hit in both legs with *incendiary* machine-gun bullets from the ME-110, and phosphorous had started to attack the bones in both legs.)

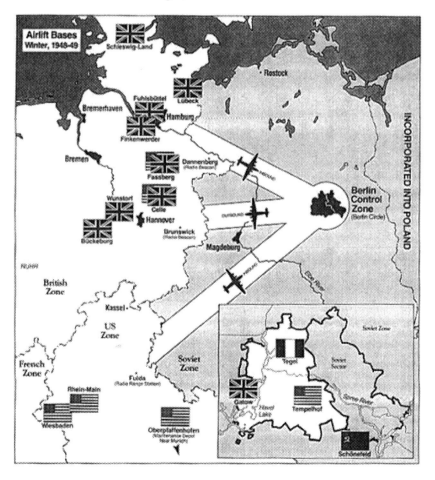

Courtesy of Dennis M. Giangreco, author, *Airbridge to Berlin*

Chapter 11: Norman Loses His Legs

In late 1949, Norman was finally hospitalized and had to have one leg amputated below the knee. He had great resolve that this would not slow him down, and indeed, it did not. Norman was fitted with an artificial limb and learned to walk again, without assistance. Losing the leg had ended his airline flying career, but not his spirit. Norman had been inspired by Sir Winston Churchill's speech, during World War II, to "never give up." Therefore he enrolled in a rehab training program and learned to become an excellent carpenter and cabinet maker.

In the early 1950s, Norman was continuing to get medical treatments for his legs. His remaining leg was still giving him some problems, but Norman kept on learning and working at his new trade. During these years, he continued to increase his skills at woodworking and cabinet making.

In 1954, Norman finally had to have his other leg amputated below the knee. He knew the drill, as he had gone through this before. Norman said that there were many veterans from World War II that had lost limbs, and the British government was hard pressed to help them in their plight. Now Norman was a *double leg amputee*, but learned, in rehab, to get around on his two wooden legs with hardly a limp. He walked the rest of his life without crutches or canes. He really was an amazing man.

In the late 1950s, during his rehabilitation, Norman worked for a tourist company that specialized in tours to Spain. He made hotel and travel arrangements for clients of the company, using his outgoing personality and fluency in the Spanish language.

In the 1960s, Norman expanded his woodworking business to building theater and, later, movie sets. He started out building room sets and became so adept at this that he was called upon quite often to build more elaborate productions. On occasion, his wife, Edna, was able to help Norman get jobs with some of the movie companies, because she was working as an accountant for one at the time, and had good connections.

Norman practiced his cabinet-maker trade, working for a large company in London. On one occasion they had a contract to build a huge, above-ground water tank for a boat show exhibition in Central London's Earls' Court. This tank was to accommodate a large boat. Nothing like it had ever been attempted there before and Norman had the design responsibility. He held his breath, and worried a lot, but the design worked well and the boat and tank held together for the full ten days of the show. Norman decided that he would go into his own business after that. Norman enjoyed his independence too much to continue to work for anyone else. Thereafter, Norman operated his own business, at his own pace. Over the years, he also built sets for exhibitions and advertising companies that did print ads and TV commercials.

In late 1977, the most famous film that Norman designed sets for was *Wild Geese*, which starred Richard Burton, Richard Harris, Roger Moore and Stewart Granger, among others. This was an epic tale of a small group of hand-picked SAS-type commandos, who were flown into Africa to rescue the imprisoned former president of a third-world country. Norman's excellent work on the movie sets was highly praised.

Chapter 12: Back to Chile

In early 1979, Norman was getting restless and ready for a new adventure. He decided to go back to Chile and look into setting up some sort of a business. He made a trial run by actually ordering some import clothing items and neckties from the Orient. He also ordered some novelty items from England. The new merchandise was to arrive in Chile during October of 1979, in time for Christmas sales. Norman leased a small store in the upscale Santiago suburb of Las Condes and went about updating the interior of the shop. He then had his wife, Edna, and daughter, Helen, leave England and move to Santiago to operate the new boutique. Neither was proficient in the Spanish language at the time, but found that about half their customers in the area spoke English. Norman handled the rest, and took care of public relations. When the imported goods arrived at Chilean customs, Norman discovered that over half of the goods had been stolen. This left them short of merchandise for the Christmas sales season, but Norman was not deterred. The ties that he had ordered from the Orient had bright colors that would "knock your socks off" and were not the traditional ones that were popular in Chile at the time. He must have set a new trend, because they were all sold out in a short time.

Norman was always itching for a new project and something to keep his active mind busy, so he decided to refinish the wood floor of the shop. After closing one evening, he got a large can of "quick drying" varnish and proceeded to coat the floor of the boutique. The next morning when Edna and Helen opened the door, their shoes immediately stuck to the varnished floor, which was definitely *not* dry. Arriving customers were met with profuse apologies. Norman arrived an hour later and was met with looks that could kill. He said, "Don't worry, I'll fix it." His answer, CARPET. End of story.

They found out that sales were very seasonal in Chile. Norman closed the shop at the end of December 1979, realizing that they needed something that would sustain them year round. Now they were ready for the next of Norman's great adventures...the English pub.

Chapter 13: The English Pub

In 1980 Norman and Edna opened the Cross Keys Pub in the Las Condes suburb of Santiago. It was an excellent location, just outside a major subway entrance. Their pub business prospered. The Cross Keys Pub in Santiago became the unofficial "watering hole" of a lot of English-speaking people in the city. Many people from the various embassies and consulates would congregate there, along with airline crews and others. There was always an interesting cast of characters gathered at the Cross Keys, and Norman and Edna were the perfect host and hostess. Norman had built most of the furnishings for the pub himself. There didn't seem to be anything that Norman couldn't do, once he set his mind to it. This was amazing in that though Norman was a double leg amputee, he seemed to have the stamina and drive of a much younger man.

On a more serious note, there was an incident which occurred at the Cross Keys Pub in 1982, during the time of the Falkland Islands War (Britain vs. Argentina). The Argentine government had laid claim to these islands for many years and called them the "Malvinas Islands." The pub was very busy during this period of time and Norman had hired extra staff to help them cope with the crowds. Norman struck up conversations with a number of husky English chaps who were frequenting the pub while the war was raging. He found out that they were all in "communications" and their homes in the U.K. were in the Hereford area. This is the headquarters area of the elite Special Air Services (SAS), Special Forces. Norman and Edna then realized that the English group occupying the "darts room" at the pub were SAS. The men had been told to stay away from the British embassy and keep a low profile. Sometimes they would disappear for a week or so, but then return. Days spent tossing darts at the Cross Keys Pub was not bad duty for these men and they appreciated the opportunity while it lasted. Norman would go by the British embassy "com center" each morning to get the latest update on the progress of the war, and then return to post the news dispatch on the bulletin board in the darts room.

One afternoon just after lunch, Edna was cleaning the bar and getting ready for a few hours' break, after which Norman would take over and get everything ready for the evening session. Norman was in the darts room, chatting away, when suddenly, in walked three hefty young men with a large Alsatian dog on a short leash. They asked Edna if this was the "English pub" and then brandished a knife which they stuck into the top surface of the bar. They said that they wanted to speak to the owner. Edna sent one of the waiters to fetch Norman from the darts room. He caught on very quickly that something was very much amiss as the three men questioned him about the "Malvinas" war and his position on it. They informed him that they were from Argentina and said that the best thing he could do would be to remove all the staff (who were then having their lunch) to a safe place, because they were going to smash the pub to pieces.

Norman didn't bat an eye, except to give Edna a sly wink, and said, "I think that before you proceed to do this, you should come with me, please....there is something that I would like to show you." He led them into the darts room where about six strapping SAS men were playing darts. He said to the Argentineans, in Spanish, "These chaps are British SAS Special Forces commandos and one word from me and they will pulverize you. I suggest that if you want to live to see another day, you get the hell out of here and take your mutt with you." Norman then pulled the knife out of the bar counter, handed it to the Argentineans and said, "Take your puny protection with you and don't ever show your ugly faces around here again." They were gone in a flash and the pub was never troubled again. Norman gave all the staff a glass of brandy, as some were very shaken, but in next to no time, things were back to normal. Everyone raised their glasses high in a toast to Norman.

The Cross Keys Pub prospered greatly in 1982 and 1983. Eastern Air Lines crews were now flying to Santiago and learning about the pub and the warm hospitality of Norman and Edna. Business grew so much that they expanded the pub by leasing two of the adjoining shops, in which Norman constructed new booths and furnishings. The crowds packed the place, until the end of December 1983, when they unfortunately lost their lease on the space in the two adjoining shops. The crowds couldn't be accommodated in the original pub space so they decided to sell, at a good price, and move.

The Cross Keys Pub, Santiago, Chile, 1983.
Barmaid Chris, me and Norman's wife, Edna.

Chapter 14: Starting Over... Again

In 1984, they moved to Algaroba, Chile, near the Pacific coast, and opened another pub. Again, Norman went to work and built most of the pub furnishings. It had only been open for about six months, when a severe earthquake in the area wiped it out. There had been no earthquake insurance available and Norman and Edna lost the pub, but not their spirit. They decided to move back to Santiago to start over.

In 1985, they opened the new Cross Keys Pub, in a spot about a block from their old location in Santiago. Many of their old customers and friends were delighted to be able to enjoy their hospitality once again. By this time their daughter, Helen, had left her flight attendant job with Eastern Air Lines and was looking for a secretarial job. She had graduated from a secretarial school in England just prior to moving to Chile and was proficient in typing and shorthand. At the new Cross Keys Pub one evening, the Australian ambassador overheard Norman saying that Helen needed a job and he promptly offered her one at his embassy. She was very proficient in the Spanish language and had worked there for about six months when the British ambassador heard about it. He asked her why she was working for the "Aussies" when she could be working for him. He called the Australian ambassador and said that he was stealing her to work for himself at the British embassy. Helen went to work for the British ambassador and did so well that she became vice counsel, later, when the vacancy came available.

In 1986, Norman and Edna, after having rebuilt the business, decided to sell the new Cross Keys Pub and retire to Viña del Mar, on the Pacific coast. Viña is a beautiful small city located just north of Valparaiso and about 40 miles west of Santiago. The climate and area is very much like the San Diego to La Jolla, California, region, with beautiful beaches and mountains.

In the late 1980s, I was able to visit with Norman and Edna on a number of occasions, while on my airline layovers in Santiago. They had a nicely furnished apartment, with a garden, in Viña del Mar. Their apartment was within walking distance of many cafes, pubs and grocers. When travel to further distant spots in the city was required, local taxis were available. There was excellent bus service to and from Santiago, that took less than an hour of travel time. This was on a very scenic route to the Pacific Ocean, and their new retirement home in "Viña."

Norman took on another project in these "retirement" years and furnished a new pub in Santiago, called the "Golden Bell," for his son, Ron. It was located in a busy suburb of Santiago, and was operated by Ron and his wife. Ron had been diagnosed with juvenile diabetes at the age of six and was a bit fragile. Norman wanted to set up his son in a business that he was familiar with. He knew that he could help Ron quite a bit, since he knew all the wholesalers and vendors in the area.

NOTE: In June of 1988, I took early retirement from Eastern Air Lines. A corporate raider had taken over the company, and I didn't see much future for my airline. I was able to travel to Chile on Eastern until it was put in bankruptcy in 1990, so was still able to visit some with Norman and Edna. Norman was also able to visit with my wife and me in Miami, several times, while passing through on trips back to the U.K. Norman was a big hit at our meetings of the "Quiet Birdmen," (a pilot's social organization that had been formed after WW I). At the meetings, he met and got reacquainted with some of the old-time aviation pioneers from Panagra and Braniff airlines. Norman boggled their minds with the story, among others, of being hired to fly for Panagra at the age of 16. He also attended some of our Retired Eastern Pilots Association (REPA) luncheons with me, and was always a big hit.

Back of card   Front of card

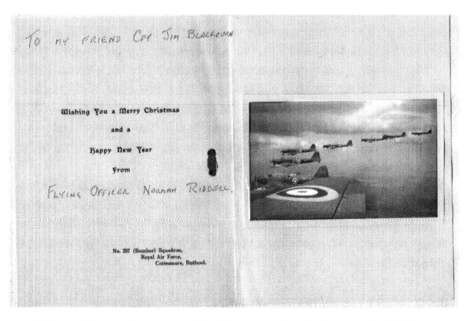

This card was given to me by Norman, for Christmas, in 1988…
(50 years after it was made!)

Chapter 15: Norman's "New Legs"

In January 1991, Norman's daughter, Helen, decided to move to the U.K. after her young daughter had become sick with a respiratory illness. The air in Santiago is very bad most of the time, due to a lot of diesel engine exhaust fumes being trapped in the valley where the city is located. Doctors told Helen that her young daughter may not survive unless they moved to a location with cleaner air. She gave up her job at the British embassy and, with her husband and daughter, moved back to the U.K.

Norman was having some problems with his old wooden legs at this time, so he and Edna decided that it was time for them, like their daughter, to move back to the U.K. Norman could get some good medical treatment there, and "new legs," as he put it. He was always able to not only keep up with, but out-walk me on our strolls around "Viña." After the move to England in February of 1991, Norman's new doctors told him that his old wooden legs belonged in a medical museum. They treated Norman with great respect, as one of the few surviving heroes of the Battle of Britain. They fitted him with the latest leg prostheses, which allowed Norman much greater comfort while walking. He really appreciated his "new legs."

In May of 1991, tragedy struck the family. Norman and Edna's youngest son, Michael, who had been living in England, was killed in a freak automobile accident. This was just three months after their return to the U.K. Norman and his surviving family grieved, but pressed on.

Chapter 16: Norman, the Band Manager

Norman would try to make at least one trip a year, to revisit Chile.

In January of 1994, Norman took Edna along with him on a trip back to Santiago. They visited with many friends and family. During this visit, Norman took Edna and her sister-in-law, Audrey, on a getaway holiday to the south of Chile. Norman wanted to relax in the resort town of Villa Rica for a week while Edna and Audrey went shopping and sightseeing further south in the city of Puerto Montt.

After a day or two of relaxation, Norman got restless again and struck up a conversation with a sad man in a local pub. This chap was a member of a six-man band and had a perplexing problem. This was the situation. The band had been hired to play at a large local tourist hotel, but was being treated like dirt. The hotel management was not living up to the terms of the band's contract. The band members had been put into the worst accommodations at the hotel, three to a room. They were provided with inferior meals, and had to eat with the hotel kitchen staff. Even some of their pay was being withheld from them to cover "incidentals." This was all in violation of their contract and their band manager (agent) had done nothing to remedy the situation. Norman thought that the hotel management may have paid him a bribe to leave, because he no longer was looking out for the band's best interest. Then it was Norman to the rescue! He asked the band members if they wanted him to become their new manager. It was a unanimous decision for Norman to take over, so he located a local barrister who drew up the proper legal papers and contract. Norman then confronted the hotel management with the violations of their contract with the band members. He also said that they would *not* perform at the hotel until the pay and living conditions of their contract had been met. If the hotel still refused to honor the contract, they would be sued. Since this was the high summer season and the band had been hired to perform for another nine weeks, through the month of March, the hotel manager found himself between a rock and a hard place and reluctantly complied. Norman even got them a raise in pay, decent rooms at the hotel and meals in the hotel dining room. The band members insisted that Norman take a manager fee for his time and trouble. Norman let the hotel management know that he, as the band's manager, would be monitoring their compliance with the contract through its completion in March. The band members praised Norman for his help and compassion in their plight. For Norman it was just another adventure, and all in a few day's work.

During the 1990s, Norman kept busy with various projects and business enterprises. He also made some visits to RAF stations and air shows. Norman still returned to Santiago for yearly visits to see his son, Ron, and some of his old friends. In January 1996, another family tragedy occurred. Ron finally died from complications of his juvenile diabetes ailment.

On Norman's trips en route back and forth to Chile, he would visit with my wife and me and some of his other Eastern Air Lines friends in Florida. On one visit, Norman used his carpentry skills in helping another retired Eastern captain. He spent nearly a month building cabinets and interior work at his home, which was located at a private airport north of Cocoa Beach, Florida. My wife and I got to visit with Norman and his friends several times while he was there.

In 1998, Norman had entered into a partnership with a small U.K. firm engaged in the purchase and sale of light aircraft. Norman was now 82 years young, and when asked, "When was the last time you flew?" replied, "Oh, it was May of this year. I wanted to try out one of our latest acquisitions."

Chapter 17: U.K. Days

In October of 1999, my wife, Lynn, and I flew over to the U.K. to spend a week with Norman and Edna. We rented a car at Heathrow Airport and drove north (on the wrong side of the road, for us) to the small village of Little Wymondley, near Hitchin in Hartfordshire. Norman and Edna had a nice apartment there and their daughter, Helen, and her family lived nearby. We had a wonderful time during our week of adventures.

Norman and I drove on day trips to some world-class aircraft museums and airfields, while Lynn and Edna went shopping and sightseeing.

Norman posing as the "Ace of the Base" with a
Hawker Hurricane fighter at Duxford in October of 1999.

NOTE: Norman and I visited the Imperial War Museum at Duxford, England. This was also the location of the American Air Museum in Britain. A few years before, I had become a founding member of the museum, and was able to sponsor a remembrance of Norman's RAF service in the Battle of Britain. We picked a great day to visit Duxford, as they were practicing for an air show the following day. We got to see Hawker Hurricanes, Spitfires, P-51s, P-47s and others in aerobatic flight. Norman loved hearing the sound of those old "Merlin" engines. Some of the old hangars at Duxford dated back to World War I, and the runways were still surfaced with grass. Norman said that his father had flown out of Duxford Airdrome in World War I. We also visited the huge new hangar building, which houses the American Air Museum in Britain collection. Another day, Norman directed me on a drive to the Old Warden Aerodrome, where the noted Shuttleworth Collection of famous and antique aircraft is on display. Many of the planes at Old Warden were flyable and would be put through their paces for the public, on some weekends.

After a day of driving, Norman and I would "lift a pint or two" in one of the area pubs. On occasion, Norman would relate some of his wartime adventures to me, but most of our talk was about flying and our bygone days in Chile.

Then it was back to pick up our wives and have dinner at a local pub near their home. We had a great visit and stored away many wonderful memories.

In June 2002, Norman took Edna to Spain to show her some of the places he had been during the bad days of 1936. He pointed out some of the bullet holes that still remained in some of the walls in Madrid, where he had been fighting during the Spanish Civil War.

We would communicate by Christmas cards and phone calls every year.

Last photo of Norman with Edna, Christmas 2003.

The last time I spoke with Norman by phone was in December of 2003. Edna had told me that he was very ill with cancer, but still getting around on his own. Norman died on February 2, 2004, and his memory and wit were sharp until the very end.

I'll never cease to remember this fine gentleman...the most unforgettable character I've ever met.

—Captain Jim Blackburn

In memorium

RAF Wing Commander,

Norman Charles Stewart Riddell

DSO, DFC, Croix de Guerre

1916 – 2004

*"Never in the course of human conflict
was so much owed by so many to so few"*

Sir Winston Churchill

Printed in the United Kingdom
by Lightning Source UK Ltd.
122356UK00001B/436/A